FUELING CURIOSITY

THE ABCS OF REFINING

A STEM ABC BOOK ABOUT REFINERIES, FUELS, AND HOW THE WORLD MOVES

BY TAYLOR BLOOMQUIST

FUELING CURIOSITY PRESS

Published by Fueling Curiosity Press
ISBN: 979-8-234-00820-6

For information, contact:
hello@fuelingcuriosity.com
fuelingcuriosity.com

For the dedicated employees at refineries everywhere: may your friends and family understand what you do every day a little bit more!

For Theodore and Vivienne: may your curiosity always be fueled!

A Note to Parents & Educators:

Do you ever stop and consider the complex systems required to support society today? The cars on the road, the planes in the sky, and most of the everyday items around your home all rely on a hidden world of incredible engineering.

This book is a window into that world.
A refinery is essentially a giant, high-tech recycling center and chemistry lab combined. Refineries take raw, heavy crude oil straight from the earth and carefully separate, clean, and reshape it into the precise fuels and materials that keep our modern society moving.

Fueling Curiosity's goal is to introduce young minds and parents alike to the foundational concepts of engineering, refining, and industrial logistics. While the illustrations are friendly and simplified, the science behind them is real. Use this book as a starting point to ask questions, explore how things work, and spark a lifelong interest in STEM!

Aa

A is for Acid

Refineries use powerful acids to help smaller fuel components stick together and make clean, valuable gasoline. This chemical process is called alkylation, which uses acids like HF or Sulfuric Acid.

Bb

B is for Barge

Barges are floating oil carriers pulled by tugboats to move crude and fuel along rivers and coasts. By connecting cities and customers, barges help make fuel cheaper and more available for everyone.

Cc

C is for Crude

Crude oil is extracted from underground. Crude travels around the world to be processed at refineries, bringing all kinds of materials used in cars, planes, and homes!

Dd

D is for Distillation

Distillation is the refinery's way of sorting crude oil into different parts. When crude is heated, light pieces rise and heavy pieces sink, so each product ends up in the right place.

Ee

E is for Emissions

Refineries work hard to keep the air safe for nearby communities. Special sensors watch the air around the refinery all the time. They can spot tiny amounts of chemicals, even as small as one drop in a swimming pool.

Ff

F is for FCC (Fluid Catalytic Cracker)

The FCC is the refinery's molecule splitter. The unit uses tons of tiny helper beads, called catalyst. These beads break big oil molecules into smaller, useful pieces for gasoline and diesel.

Gg

G is for Gasoline

Gasoline is not just one liquid. Gas is a carefully tuned mix of many refinery products. Refineries blend and test the mixture to ensure engines run smoothly and clean.

Hh

H is for Hydrotreating

Hydrotreating uses hydrogen (H) to clean fuel. The hydrogen grabs sulfur (S) and other unwanted contaminants and carries them away. Removing sulfur helps the fuel burn cleaner.

Ii

I is for Instrumentation

Instrumentation is the refinery's monitoring system. Sensors watch temperature, pressure, flow, and levels, and send that information to the control room, which helps operators keep units running at their best.

Jj

J is for Jet Fuel

Jet fuel is special because it is made to work high in the sky, where it is very cold. The fuel will not freeze or leave behind residue, helping engines stay healthy for a long time.

❄ *Cold ready*

Kk

K is for Kerosene

Kerosene lamps provided light long before electric lights were available. Refineries now refine kerosene to be cleaner for jets and even rockets headed to space!

Ll

L is for Loading Rack

A loading rack is where tanker trucks fill up safely and quickly. The racks provide gasoline and diesel to all the local stations, regardless of their brand.

Mm

M is for Maintenance

Maintenance keeps equipment safe and dependable. The reliability team repairs, inspects, and replaces pumps or valves so the refinery can operate without unplanned interruptions.

Nn

N is for Natural Gas

Natural gas is a clean-burning fuel used to provide heat for refineries, just like it heats stoves in our homes. The gas's steady flame helps heat equipment and oil efficiently.

Home Cooking

Process Heat

Oo

O is for Operator

Operators are the pilots of the refinery adjusting temperatures, pressures, and flows as necessary. Console operators watch screens and trends, while field operators work directly in the unit.

Pp

P is for Pipeline

Pipelines move huge amounts of fuel underground, all day and night. One pipeline can replace hundreds of trucks per day, improving roadway traffic and transportation safety.

Qq

Q is for Quality Control

Refinery labs check the materials coming in, the products going out, and everything in between. These tests make sure everything meets the right standards, so the fuel works the way it should.

Rr

R is for Reformer

The reformer makes gasoline stronger. The unit takes straight hydrocarbons and rearranges them into better shapes, like rings, so engines can run predictably. Along the way, this also releases hydrogen, which helps other refinery units.

Higher octane

Ss

S is for Steam

Steam is used for heating and cleaning out equipment. The large white plume that is visible is typically steam, which is simply water vapor condensing. Clouds become visible when the warm vapor encounters cooler air.

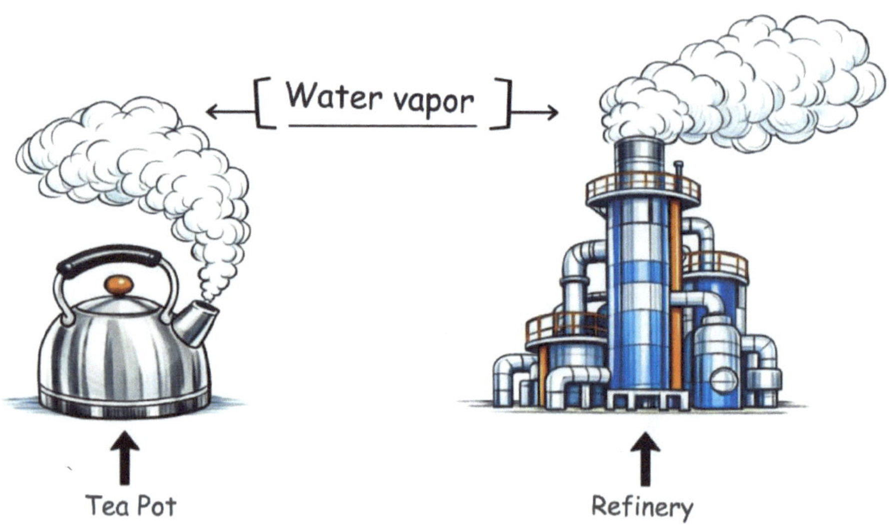

Tt

T is for Turnaround

A turnaround is a planned shutdown of the refinery for a deep clean. Crews open equipment, inspect it, clean it, and repair it. They work day and night so the unit can restart on time and run for several years.

Uu

U is for Ultra Low Sulfur Diesel

Ultra low sulfur diesel is made by removing most of the sulfur that contributes to pollution.
This cleaner diesel powers the majority of heavy machinery due to its dense energy content.

Cleaner diesel

Vv

V is for Vacuum Tower

A vacuum tower lowers the pressure and lets heavy oil float like vapors at lower temperatures. This process helps separate valuable material gently without overheating and burning it.

Ww

W is for Wastewater

After water is used by a refinery, it is meticulously tested and cleaned before it leaves the site. Multiple steps remove oil, solids, and impurities so the water can be safely returned or reused.

Xx

X is for X Ray

X rays help inspectors check the thickness of pipes without stopping everything to look inside. The images show if the pipes are safe to keep using or need to be repaired.

Yy

Y is for Yield

Yield means getting the most useful products from each barrel of oil. Refineries work hard to turn almost everything into fuel or other helpful materials without wasting anything.

Every last drop counts!

Zz

Z is for Zeolite

Zeolite is a special kind of catalyst material with tiny, shape-matched tunnels with unique reactive spots. Refineries use many different catalysts to make the right products.

Only the right molecules fit.

Beyond the ABCs: Refining Explained

This section is written for parents, educators, and curious adults. It expands on the ABCs by explaining how real refineries operate, why tradeoffs exist, and how decisions affect production, cost, and fuel quality.

A – Acid (Alkylation)

Alkylation units using hydrofluoric (HF) acid are designed with multiple layers of containment, monitoring, and emergency protection because HF requires exceptional care. These units operate under some of the strictest safety standards in the refinery, producing one of the cleanest and most precisely controlled gasoline components available.

B – Barge

Barges allow refineries to move very large volumes with minimal fuel use per barrel transported compared to trucking. Their schedules must be carefully coordinated with tides, locks, and terminal availability. Small delays on waterways from wind, rain, or fog can ripple into refinery inventory planning. If barges are forced to wait, they charge a fee called demurrage, which can cost over $10,000 per day.

C – Crude

Not all crude oil is the same, and each type behaves differently once it enters a refinery. Differences in density, sulfur, and metal content change how units must be operated. Refineries often blend multiple crudes from across the globe to create a feed that fits their equipment and product goals.

D – Distillation

Distillation separates crude by boiling range without changing the molecules themselves. Poor separation at this stage increases the workload and cost of downstream units. As an example, precise temperature control allows refineries to push the lightest part of diesel into gasoline or the heaviest part of gasoline into diesel, depending on the market.

E – Emissions

Many emissions controls focus on preventing leaks rather than detecting dirty exhaust. Proper sealing, maintenance, and operating discipline reduce releases before they occur. Leak Detection and Repair (LDAR) programs check fittings and equipment on daily routes using advanced infrared cameras and other technologies to identify and repair leaks of gases and vapors that are invisible to the naked eye.

F – FCC (Fluid Catalytic Cracker)

FCC units can change temperatures to increase or decrease how much they crack their heavy oils. Lowering temperatures can increase diesel production and higher temperatures increase gasoline. Refiners watch the market closely to fill the greatest need.

G – Gasoline

Gasoline recipes change throughout the year to match seasons and regulations. Vapor pressure, how easily gasoline can turn into vapor, limits are adjusted throughout the year to control emissions and improve drivability. In the winter, a higher vapor pressure limit allows refineries to blend more butane and make gasoline cheaper.

H – Hydrotreating

The sulfur removed by hydrotreating is processed by the SRU into molten sulfur, which solidifies into bright yellow powder. This material is then sold into petrochemical markets or even for making fertilizer. Hydrotreating also removes nitrogen and helps saturate aromatic molecules, improving fuel stability and emissions performance.

I – Instrumentation

Critical measurements often use multiple redundant instruments for same temperature or pressure. Comparing the readings helps detect sensor failures before they affect operations. Reliable data is essential for safe control decisions. When all else fails, measurements are verified in the field by operators.

J – Jet Fuel

Jet fuel's ability to resist cold temperatures affects diesel markets as well. During winter seasons, refiners may blend jet-range material into diesel to improve cold-weather performance for trucks and heavy equipment. This also drives export markets as refiners with ocean access can sell into warmer climate conditions if they do not have jet to blend economically.

K – Kerosene

Not all kerosene can become rocket fuel. Rocket-grade kerosene requires exceptionally tight control over sulfur, aromatics, trace metals, and thermal stability because even tiny impurities can damage engines operating under extreme temperatures and pressures. As a result, only a small number of refineries are capable of producing kerosene that consistently meets these specifications. Even fewer are approved suppliers.

L – Loading Rack

Loading racks use automated controls to prevent overfills and misrouting. Flow rates, valves, and grounding systems work together for safety. These systems allow fast, accurate product delivery. While the base fuel is all the same, fueling stations do add brand specific chemicals for engine performance and cleaning.

M – Maintenance

Modern maintenance relies heavily on condition monitoring rather than fixed schedules. AI continues to play a deeper role in discovering and predicting the earliest indication of issues. Early intervention reduces costs from full violent mechanical breakdowns to minor and controlled part replacements.

N – Natural Gas

Natural gas is the primary heat source for most refinery furnaces and boilers, making it essential to daily operations. Its availability directly limits how much crude a refinery can process. During periods of high regional demand, especially in cold weather, natural gas supply can tighten, forcing refiners to reduce rates or re-optimize operations.

O – Operator

Field operators are present at the refinery 24/7 working in shifts. Console operators, however, can be located hundreds or even thousands of miles away! Information transfer between shifts is one of the most critical operator tasks. They interpret trends, not just alarms. Subtle changes can indicate developing issues long before limits are reached. Experience plays a critical role in safe operation.

P – Pipeline

Pipelines often move different products in sequence through the same line. Operators carefully manage the interface between batches to prevent contamination. This requires precise scheduling and control. Also, advanced leak detection technologies can identify a pipeline leak within minutes, even when the pipeline stretches hundreds of miles.

Q – Quality Control

Refinery labs run nearly 2000 tests per day on various streams moving between the units and across the fenceline. The results directly influence shipping decisions, and delays in results can hold products in tanks and disrupt logistics. Fast, accurate testing supports smooth refinery operations. Many times, refineries also must have products tested by an independent lab before shipping.

R – Reformer

Reformers intentionally trade volume for higher octane quality. Releasing hydrogen causes volume shrinkage, which is planned into refinery economics. Reformers are also a major contributor to petrochemical uses such as plastics and solvents. Petrochemical and gasoline markets compete for reformer products.

S – Steam

Steam production, use, and losses are closely tracked. Leaks, poor insulation, or incorrect pressure selection can waste significant energy. Improving steam balance is one of the fastest ways to reduce operating costs. It can also be injected directly into the process to help carry lighter hydrocarbons away from heavier ones.

T – Turnaround

Turnarounds are planned years in advance and involve thousands of coordinated tasks. Every day the units are down can cost millions, so every hour is tightly monitored and scheduled. Scope control is critical to prevent schedule overruns. Successful turnarounds set the foundation for years of safe operation.

U – Ultra Low Sulfur Diesel

ULSD production not only removes sulfur but also adds hydrogen to molecules called aromatics. This causes the molecules to expand and increase in volume. While this is economical, it increases hydrogen consumption across the refinery. Hydrogen availability can limit diesel output. Balancing diesel demand with hydrogen supply is essential and can drive catalyst selection.

V – Vacuum Tower

Vacuum towers help upgrade asphalt and rely on airtight systems to maintain low pressure. Most refineries create the vacuum using high-velocity steam ejectors that rush by a small opening and drag air and vapors out of the system. Air leaks cause pressure increases and reduces separation efficiency. Maintaining vacuum integrity protects product quality.

W – Wastewater

Wastewater systems must handle both routine flows and sudden upsets. Heavy rainfall or process changes can stress capacity. Treatment limits can constrain refinery rates. This process even uses bacteria systems to digest trace oils, and refineries take great care to make sure the "bugs" are happy and healthy.

X – X-Ray

X-rays inspections gradually span the whole plant, conducted at night to reduce work disruption, minimizing the risk of an unexpected failure. If thin areas are identified, refineries may install clamps around the pipe to provide a second layer of pressure control to continue operation until a controlled repair can be planned. Areas near welds, elbows, and low points are prioritized for X-ray inspection because changes in flow, chemistry, and temperature tend to accelerate metal loss.

Y – Yield

Yield improvement is rarely free of tradeoffs. Maximizing one product can reduce flexibility elsewhere. Refinery optimization balances value, cost, and operability. Refiners use complex models called LPs to determine the most economical configuration of yields. Loss control teams investigate any discrepancies, helping to identify leaks, measurement errors, or process inefficiencies that can impact overall refinery performance.

Z – Zeolite (Catalyst)

Zeolites are one of several catalyst types used in refineries. Other catalysts are designed to promote chemical reactions by adding or removing hydrogen, often using precious metals like platinum to speed those reactions. Each catalyst type is tested and chosen for a specific job, and the selection strongly influences fuel quality, yield, and operating conditions.

From the Author:

Fuel is something most of us use every day without ever seeing where it comes from. Behind every gallon is a careful balance of science, engineering, logistics, and people working together to keep the world moving safely and reliably.

Refineries are not just places where fuel is made; they are systems rewarding curiosity, discipline, and thoughtful decision-making with significant opportunity for innovation.

If this book helped you see the hidden world of energy a little more clearly, then it has done its job.

The next step is asking more questions, noticing how things work, and staying curious about the systems we depend on every day. Curiosity, after all, is what fuels progress.

Please reach out to hello@fuelingcuriosity.com or visit fuelingcuriosity.com to learn more about the flow of crude to consumer.

9 798234 008206